Counting Birds

D.L. MCCUIN

The art in this book was created with images of birds extracted from photographs.

ISBN-10: 1-52-371095-0 / ISBN-13: 978-1523710959 (pbk)

McCuin, D.L.
    Counting Birds/D.L. McCuin — 1st ed.

    Summary: Count from one to ten while becoming familiar with the world of birds.

    [1. Counting — Juvenile literature. 2. Birds — Juvenile literature.]   Title.
2015
[E]

For my husband

1

one

Greater Roadrunner

2

two

Northern Cardinal

3

three

Eastern Bluebird

4

four

Spotted Towhee

5

five

Vermilion Flycatcher

6

six

Black-headed Grosbeak

7

seven

Tree Swallow

8

eight

Eastern Phoebe

9

nine

Painted Bunting

10

ten

White-eyed Vireo

Did you say, "wren?"
No?
Ohhh, ten.

If you're looking for ten,
turn the page and begin.

Bewick's Wren

10

ten

9

nine

8

eight

7

seven

6

six

5

five

4

four

3

three

2

two

1

one

And you're done!

Oops! Maybe not. Go to the next page for more fun!

The Count Up

1. How many **kinds** of birds are standing on an object?

2. How many **kinds** of birds have red on them? (Hint: look carefully. There are a couple that have only a little red.)

3. How many **kinds** of birds have blue on them?

4. How many birds have their mouths open? (Hint: count all of them, even the duplicates.)

5. Challenge: How many White-eyed Vireos are there in the book?

Don't forget to count me!

www.ingramcontent.com/pod-product-compliance
Lightning Source LLC
Chambersburg PA
CBHW050428180526
45159CB00005B/2453